Swarm Traps
and
Bait Hives

THE EASY WAY TO GET BEES FOR FREE

McCartney Taylor

Contents

Foreword

What Is in It for You?

*P*hilosophy is the most important of all the sciences.

As an engineer, I grudgingly must admit this. Through gritted teeth. Really. The underlying "why" of all of our actions is the most critical part of our lives.

In this case, I ask you "Why are you beekeeping?" and "Why do you want to catch swarms?"

The answers to these questions will define you as a beekeeper. So pause and think on them longer than you normally would.

1.

Why Trap Swarms?

*I*n a nutshell, beekeepers set up bait hives and swarm traps for free bees.

After all, I'm a lot like you: I'm a beekeeper and want to increase my hive count cheaply. Swarm traps give me free bees, a lot of free bees. In fact, sometimes I have too many hives, and occasionally I sell them off for spare cash. It works for me, for my fellow beekeepers and it can work for you, too.

Catching them is an easy, biologically driven process — nothing fancy, nothing expensive, nothing dangerous. In fact, it is downright fun. I think of it as fishing, just like setting out trotlines for catfish and checking them weekly or monthly. Note that the name "swarm trap" is a misnomer. The bees are free to come and go. A better name would have been "decoy hive box," but swarm trap is the name that caught on. Beekeepers also use the term bait hive.

Swarm trapping works for me, and it will work for you, too.

2.

Trees

*T*he rule of thumb is this: "Where there are trees, there are bees."

Any forest in any setting will give bees the shelter they need.

Where there are trees, there are bees.

Bees also consider buildings trees. In the United States, building density is so high in urban areas that bees consider them forest. So there are bees in cities in the Southwest where there are no trees.

If you live on a treeless island somewhere off the eastern coast of Canada, you may not have any bees. Sorry. We recommend that you move, as it has to really be a bummer to live there, whether you are a beekeepers or not.

3.
Biology of Swarming

One purpose of all living things is to go forth and multiply.

Bees are no different. It doesn't matter if they are Apis M. Meliffera, Apis Dorsata or Apis M. Syriaca. They all want to procreate and keep their species propagating. The difference is often amongst the species of Apis by how aggressive they are in propagating.

Bees multiply by swarming. The biology of this phenomenon is quite interesting. There are two main causes of swarming:

1. Overcrowding: The hive is congested and is forced to swarm.
2. Reproductive Instinct: The hive has the innate urge to split.

The first cause is obvious: The hive broods up in spring and becomes too congested to function, so it swarms.

The second is more interesting. The entire hive, not the queen, decides that there are enough resources—honey, pollen and bees—for the host hive to survive if part of the hive leaves and enough time for a new hive to establish itself before winter. Because bees typically swarm only when they have enough resources, most swarming takes place during honeyflow. In the southern United States, honeyflow is usually in Spring and again in Fall. Fall swarms are unusual, however, and usually triggered by overcrowding, in my opinion.

There is probably a doctoral dissertation on the causes of Fall swarms waiting for some entomology graduate student.

Once the swarm impulse triggers in the hive, several things happen:

1. Queen cups get made (these are peanut-looking cups normally made on the lower half of the comb).
2. The queen deposits eggs in the cups, thereby making them officially swarm cells.
3. The brood nest gets backfilled with honey.
4. The queen slows down egg laying.
5. The workers get the queen slimmed down for flight.
6. Forager bees become scouts looking for a new home for the swarm.

A note on number three, this is often called "honey bound." There is a hot debate as to which comes first:

1. A honeyflow is so good that all cells get filled with honey and the queen has nowhere to lay so the hive swarms.
2. The hive has chosen to swarm therefore it backfills the brood nest.

I currently have no opinion on this; however, beekeepers are notorious for the tenacity of their opinions. Ask three beekeepers a question, and you'll often get five answers.

When the first queen cell gets capped, the bees who are leaving with the swarm fill up on honey, and the entire swarm leaves with the old queen. This is called the "prime swarm." They don't wait for the new queen to emerge but simply wait for the queen cells to be capped then depart. This is usually 8 days after starting cells.

When the prime swarm first leaves the donor hive, they usually land on a temporary staging area. This is often a tree branch, but they can land on anything from a park bench to a bumper of a police car.

From this staging area, scouts from the swarm go off to find a suitable home to occupy—a cavity with a good amount of darkness and a smallish defensible hole. Of course, this is the ideal, but often, they don't find the perfect home and concede to settle for less. The process of deciding what home to occupy is quite interesting—and democratic.

When a scout finds a suitable cavity, it comes back to the swarm, tells the other scouts it has found a good site. Some of the scouts check it out and will come back to pitch the site to even more scouts. Bear in mind, there are multiple scouts coming back and trying to sell the scout population on "their" site. Ultimately, when the scout count gets up to about 15 to 20 and they visit a site and it passes muster, they go back with a different message: "WE FOUND IT !". The call to flight trigger is done by the scouts issuing a slightly audible piping of around 200 Hz. Something in the hive mind accepts the quorum decision of 15 to 20 scouts, and it stops the hunt for other locations. We aren't sure why that happens—another doctoral dissertation waiting to happen.

The study on this is "Group Decision Making in Honey Bee Swarms" by Seeley, Visscher, Passino.

The cavity that scouts find ideal is very important to us, as we wish to make our trap as close to ideal as possible. We want these swarms to feel our traps are perfect.

Through another study, "Nest-site selection in honey bees" by Seeley and Buhrman we find that the ideal volume of the perfect cavity is around 40 liters (10 U.S. gallons) in size, allows almost no light in, and has an opening of 15 square cm (2.3 square in. or about a US half dollar coin). Bees will accept a volume as small as 15 liters but prefer 40 liters by far.

Another study found that Africanized honey bees will accept cavities as small as 12 liters and show little preference between 15 and 40 liters. The lesson from that is to always have your swarms traps in the 30 to 40 liter range to avoid skewing toward capturing Africanized honey bees.

This study also showed that larger swarms do not have a preference toward larger cavities, but this finding is not accepted by many beekeepers. As the author, I'm expected to know everything but I don't know about this. I'll wait for more research. I will accept the finding for now, as the authors used the scientific method to discover this. Even then, I suspect each subspecies of Apis mellifera will react a bit differently.

Back at the donor hive, the queen cells left behind eventually pop out queens. The hive decides if it is strong enough to cast out a second hive or not. If not, the queens battle it out to see who rules and who dies.

4.

Swarm Names

The first swarm, or prime swarm, that comes out is the largest and leaves with the old mated queen.

If the hive decides to swarm again, the first virgin queen to hatch swarms away with more of the hive. This second cast is called the "bull swarm." If other swarms cast off from the hive, they are merely called casts — the third cast, the fourth cast, etc.

Be ready for spring. Some of us take swarm trapping seriously.

5.

History

*B*eekeepers have used swarm traps throughout recorded history and quite likely deep into ancient times. The beekeeper is, after all, the oldest documented profession, dating back to a 7,000-year-old cave painting in Spain.

The solution of swarm traps and bait hives is not new, nor is the need of them. Most primitive cultures with beekeeping have the chronic problem of getting a swarm to enter their production hives.

The pheromone lure is new, however. The lure research brought us the knowledge that a simple cheap lure can double, triple or quadruple our catch rate. Suddenly, what was an iffy return on our investment of time and resources, became a very high return on investment. Ironically, the number of beekeepers in the developed world has shrunk phenomenally, and there are now few of us who can use this information. So congratulations on buying this book, you did the right thing.

There is old information out there from a bygone era that is no longer has relevance due to the introduction of the lure. Also, a cheaper and equivalently effective lure emerged known as lemon grass oil sometimes called LGO.

Beekeepers in Africa already knew lemon grass oil was effective and have been rubbing lemon grass inside their hives. It just took modern science to confirm its effectiveness before beekeepers in developed countries adopted the practice. But don't be too hard on the scientist. African beekeepers also believe cow manure can also bait a hive. I have doubts on that one, or maybe the cows were eating lemon grass? Again, I smell a doctoral dissertation in the works. Or maybe I smell something else.

6.

Alternatives to Traps — Catching Swarms

*B*efore the advent of swarm traps, the easiest way to get free bees was to capture a swarm on a branch. After all, they were free bees, right? Maybe not.

First, you have to know about the swarm. Someone has to call you, text you, email you, or find some other way of telling you that there is a swarm. Some of those methods have a time delay. Take email for example. That delay may mean that by the time you call back, the swarm is gone.

But assume it is still there. So you have your swarm pickup gear in the trunk of your car, as a good beekeeper should have. You drive over to the location, and when you get there the bees are gone. Even worse, my friend Roberto was turning to get his brush for moving the bees into to the box when "*WoooooooooooooOOOOSHHHHHHHHHHHHH!!!*" the swarm left right in front of him.

Well, let us rethink free. You need:

1. Time to get sources to call you about swarms.
2. Time to arrange with owner to pick them up.

3. Time and cost to drive to the swarm site and back. Don't underestimate the real cost per mile, and don't be fooled that it is just the cost of gas. The amortized wear and tear cost to running a vehicle plus gas is currently around $.52/mile according to the IRS.

4. Time to pick up the swarm.

5. Time to drive out again or time spent waiting (if you wait until dusk for all of the bees to return to the swarm).

6. Adjust missed swarm costs into the successful calls.

7. Adjust in the time and money of travel to show up and discover someone misidentified yellow jackets for bees.

Swarm traps are easy to use. Just hang them and swarms move in.

F.

Building a Swarm Trap — The Hanging Box Type

ou can build a swarm trap from scratch or skip ahead to section called "Buy an 8 frame deep".

The trap itself is simple. Now that we know the ideal cavity size is 40 liters, it becomes easy to make the trap. You have two choices at this point: either make the box from scrap wood or use a Langstroth 8 frame deep hive body.

The main design criterion is to have frames in the trap for easy transfer to a Langstroth hive. Frames also minimize stress on the hive when you move and allow us, the beekeeper, to avoid checking the trap as often.

For top bar hive beekeepers, the same logic applies. I'll discuss the top bar hive difference later.

You may already be aware that you can buy pressed cardboard flowerpot-shaped swarm traps. These are really awful. If you catch a hive, you have to get them out quickly or they build comb and you will have to do a cutout on a swarm trap. Even if you do get them out fast, they might abscond because they have no comb. These cardboard swarm traps only last for a season or two and they don't do well in hard weather. Their price is high as well. In other words, avoid cardboard traps.

Making a trap from scrap wood offers the advantage of price and the pleasure of building, not to mention a great way to recycle short boards. The dimensions of the box are fairly loose, so even a terrible woodworker like me can do it. And I did do it—terribly. Yet, those terribly sloppy boxes I built caught swarms.

PHOTO CREDIT: JP ARMSTRONG
Cardboard trap — avoid this.

Dimensions

External dimensions of the swarm trap don't matter, since the trap may be made of scrap wood of varying sources. This will throw off any nice, neat outside measurements. Bees only care about the swarm trap's internal volume.

Above: Notice the piece of wood nailed on the inside to rest the frames on.

For internal dimensions, the box is about 19 1/4" x 13" x 10". These dimensions are loose, as long as you are close and your frames fit without the frame ears falling off, you are fine. My biggest problem is that I tend to make the trap a bit too long and the frame ears fall off. My earlier traps were 20" long, and I'd nail a stick of wood on the inside of each end for the frames to rest on. It is simple, and it works.

Remember that the inside dimension will accommodate a normal Lang deep frame. If you are in another country with a different frame size, then your dimensions will be different. Adapt. These dimensions should hold 8 deep frames. If your box holds 7-10, you are fine.

Drill an entrance hole of 1 ½" to 2" in diameter *near the bottom* of the trap. Studies show bees prefer bottom entrances. Then drive a nail across the entrance to keep birds out. I drive two nails in, making an X across the entrance, to keep flying pests out. Bees simply ignore the nails.

Once the box is built, then screw a 24-inch "hangboard" on the long side of the trap. This is usually a 1 x 6 (3/4" x 5 ½") board. Then drill a half-inch hole 2 inches from the top. Be sure the hole is perfectly centered for the trap, so it doesn't hang at a wonky angle.

I recommend using a minimum of 4 screws to screw it to the body. **Do not use nails**, as this board will take the brunt of the dynamic forces on the swarm trap. It will come apart if nailed. Trust me, I tried. Learn from my mistakes: **use screws**.

Two nails keep birds from nesting in the swarm trap first. If birds get in, bees will not hive there.

13

A trap from all angles.

To hang the swarm trap, just find a good tree, drive in a nail and hang the hangboard on the nail. Done. Check the trap in a week. I'll cover more on hanging the trap later.

Buy an 8 Frame Deep

Swarm traps happen to be darn close in size to an 8 frame deep (39 l) and a 10 frame deep (42 l). So for minimal muss and fuss, I'd recommend simply buying an 8 frame deep, as it is narrower than a 10 frame deep and will hang better as a trap. Don't use a 5 frame nuc box. As mentioned earlier, studies have shown that Africanized bees will accept undersized cavities.

14

Above: The best designs are simple in nature. Below: A purchased 8 frame deep.

If you decide to build a swarm trap with an 8 frame deep, you just have to screw it together, then find spare wood for the top and bottom. Drill a 2" hole in the bottom. This avoids ruining an 8 frame deep you might use later in the season for a nuc.

Homemade traps.

Once you have built the box, you will need to hang it. This is where this design shines in its simplicity. Like I said above, all you have to do is take a 1" x 6" board, screw it to the side of the trap and then drill a hole near the top so it can hang on a nail. This design is stable and self-righting in a strong wind.

I want to reiterate the point that no matter which design you choose, assemble the trap with screws, not nails. The trap will be on a tree for months, rocking in the wind, that makes it dynamic. Dynamic objects work nails out, but not screws. If you nail the trap together, it will slowly come apart. I'll promise you that.

The top *must* be screwed on, not nailed. Even if you used nails (you'll regret that) for the rest of it, the top *must be screws so that you can open the trap calmly.* If you have to pry the lid off, you might break new comb and definitely anger the bees.

There are a few videos of how to build a swarm trap on my beekeeping site http://learningbeekeeping.com .

Don't use a 5 frame nuc box. As mentioned earlier, studies have shown that Africanized bees will accept undersized cavities.

Bonus Nuc

Swarm traps have a nifty dual purpose, they work as a nuc, too. It already has frames and a capacity of about 8 frames. You could even overwinter a nuc in it, and have the swarm trap primed as a veteran for the spring season.

Scrap Wood Option

Bees prefer old wood over new wood. I'm not sure why, but there is a naturally occurring substance in live wood that acts as a fungicide. Perhaps this same substance bothers the bees. Or perhaps it is an innate sense that older wood rots out and will form a cavity for them. At any rate, older wood is preferred over new wood when you are making a swarm trap. Remember that new plywood may also have glue that hasn't fully out-gassed. Also, don't use chemically stained woods due to bad smells. That includes varnish or spilled oil.

When you can, grab old boards and old plywood for your bait hives throughout the year. Plywood is good for tops and bottoms, but not so good for sides. If you buy new wood, it will work. Older wood may improve captures by a few percentage points. That is my opinion with no supporting studies.

Re-purpose Broken Equipment

One final use for rotting hive bodies and old or broken frames is to repurpose them for their last leg of life as a swarm trap. After all, you have to take old frames out at some point. This is where they belong.

If you have a 10 frame deep that has a corner or two rotted out, just cut it down to a 6–8 frame body and reuse it. This approach let me chop off the rotten side and make a perfectly good trap.

Just don't be foolish. If the frame or hive body was from a diseased hive, you should burn it. Don't put known bad stuff in a swarm trap. That would just be bad Juju, so go with common sense and burn diseased equipment.

Safety

Anytime I hear about a beekeeper dying on the job, it usually involves beekeeping on a ladder. Think about being on a ladder, dealing with a stinging insect and suddenly you get stung. Is your instinct to remove the sting or hold on to the ladder and safely come down? You guessed it. Almost everyone jerks when they get stung and leaves themselves vulnerable to a tumble. Don't ever do beekeeping on a ladder unless you are geared to repel a full out bee attack. Better yet, just stay off ladders.

For swarm trapping, you shouldn't need a ladder. At most, maybe a 1-meter (3-foot) stepladder or a flipped 5 gallon bucket. Most traps should be hung less than 8 feet high, but we will address that later.

Just stay off ladders.

Top Bar Hive Option

Let's be practical. The biology of bees doesn't change if you use a top bar trap or a Langstroth trap. The same biology is true for scout bees and their need for cavity volume.

So you can just simply put top bars across the top of a swarm trap. A little adaptation will be needed on the lid to keep the bars in place, mainly two nails on each side of the trap near the top so a wire can be strung across and wire wrapped down to keep the bars from sliding. Remember that this trap is a dynamic object rocking in a tree for weeks, so it must be secured for abuse.

If you catch a swarm and it builds out a comb that won't fit in your top bar hive, just trim it to fit with your hive tool. Don't sweat the small stuff.

1. Swarm traps work for top bar hives as well.

2. Notice the nail ...

3. ... get the wire ready ....

4. ... and tie the wire down.

8.

Baiting a Swarm Trap

In the old days, swarm trappers had all kinds of stuff they thought would improve occupancy in the trap.

Some studies were done, and they decided that a trap should be 12' high and have a bit of old comb in it. The comb was the bait. It smelled of a hive. Swarms appear to recognize the value of pre-existing comb and the energy resource they will save.

The height option was valid at that time. But something new has come along. Something that renders height unimportant. Something ... **revolutionary.**

The latest revolution in swarm trap technology occurred 20 years ago with the analysis of bee pheromones. They discovered the different pheromones that bees communicate with and were able to synthetically duplicate them.

This means that we can now put a pheromones lure in a swarm trap and expect a 30% occupancy chance instead of a 5% chance in the old days with just the use of old comb.

No matter which lure you use, their effectiveness declines throughout the season. There are ways to maximize how long they last. With commercial swarm lures, there is really nothing you can do but keep them frozen until you need them and then drop them in the trap.

Lemon Grass Oil: Our Secret Weapon

Once we understood bee pheromones, we discovered that the big one for swarm trappers is the Nasanov pheromone. It smells like lemon. Consequently, beekeepers discovered lemon grass oil has the same chemical components as Nasanov pheromone. What makes is really great is the price: it is cheap.

Simple baiting: A napkin corner and 8 drops of lemon grass oil.

21

If you are using lemon grass oil, drop 8 drops on a corner of a clean napkin or paper towel. Then put that corner in a small zip-lock sandwich bag. Seal the bag and then open a small corner of the bag so the aroma to drift out. Bees have an awesome sense of smell and will smell it.

Why 8 drops? Why on a napkin? If you put the lemon grass oil on the frames, the smell dissipates too quickly. If you use 20 drops, you overwhelm with the odor of lemon. So 5 to 10 drops is the right amount. I could hypothesize that the lower your humidity level, the more drops you should use, but that would encourage yet another grad student to complete a dissertation on the topic. For us, 8 is the right number.

Every time you visit the trap to check, you can put a dab of lemon grass oil on your finger and wipe it on the front of the trap. This is much easier than taking the trap down, unscrewing the top and recharging the lure. The lure is still active, and you merely are adding a boost to the trap.

Commercial Swarm Lures

Several companies make commercial swarm lures. They are expensive, only last a few months and do the same thing lemon grass oil does. In fact, many commercial lures contain some portion of lemon grass oil. In good conscience, I won't recommend that anyone buy them. I think lemon grass oil is just as effective, and much, much cheaper.

Use Old Comb

Smells are obviously an important component to swarm traps. So, I recommend you use a piece of old comb along with lemon grass oil. Ideally, the comb is attached to the frame for use by the bees. It gives them a comb guide, too.

This comb smells of bee habitation and is a natural attractant. Black comb is recommended over white, as the black will have more of an odor. However, if all you have is yellow, use some. If you have none, that is okay, too. The lemon grass oil is doing 70% of the attraction. You are just shooting for a 10%

Old comb improves your odds of catching bees. Put some in your frames. It's attached by melting it slightly with a blowtorch.

boost. Just to clear, I'm giving these percentages as a best guess as there are no studies for me to rely on.

On the smell, the second biggest boost to your chance of success is to use a swarm trap that caught a swarm recently. The propolis and the smell of occupation permeates into the comb, the wood walls, the top and the bottom of the trap. It stays, and the scout bees smell it. My best guess estimates it give you a 20% boost. This is based on my experience and that of other trappers I've interviewed.

So the best setup currently is

1. Lemon grass oil
2. Old black comb
3. Veteran swarm trap

If you have large pieces of burr comb from other hives, cut them out and save them for your swarm traps. This is far better than just melting them down.

The trick is to attach the burr comb to frames to put in the hive. I like use a blowtorch to melt the top of the comb and quickly stick it on the top of a frame. I prepare for this technique by first getting the frames ready and then flipping them upside down to rest on their top bars and lining them up. Once I am ready, I use a blowtorch to the melt burr comb until it is flat on the top and quickly squish it onto the frame. Since the frame is upside down, the extra melted wax drips down on them, and the comb sticks quite nicely. This is the most hassle free-method I've found. If you are going to be stupid or careless, don't try this. That torch is hot, and it could burn you and catch things on fire. Show respect for open flame.

I've also taken string and a needle and sewn the comb to the frame, but this was slow and not as practical as the blowtorch method.

As a rule, avoid putting in frames with foundation. It takes longer for them to draw out, and they may have bad smells in the wax. But if frames with foundation are all you have, then use them.

Also, do not put in frames of honey as bait. It only gets robbed out and doesn't help you. As for fear of wax moth, don't worry about them getting in the bait comb. The swarm will deal with it.

QMP

There is an alternative to both lemon grass oil or commercial swarm lures. It is called Queen Manibular Pheromone, or QMP. It cannot be used with the other two, only alone. QMP is just alcohol that old queens have been dropped into thereby transfusing their scent into the alcohol.

I don't use it, as lemon grass oil is cheaper and a lot easier to get. I believe it last longer as well, so there isn't much to argue for QMP. Don't use QMP with lemon grass oil.

9.

A Note for Non-Beekeepers: The Hands-Free Beekeeping Method

*M*aybe you aren't into catching swarms for beekeeping and just want bees for pollination.

If that's the case, then you don't need to get your measurements exact on the swarm trap, and you don't need to put frames in the trap. You still need to bait the trap, and adding old comb helps. I also recommend that you hang the trap higher, out of reach and sight of neighbors and their kids. You can always claim it is a birdhouse that bees set up shop by nature's design. Essentially, your swarm trap is a means to hands-free beekeeping, without having to ever deal with bees.

Build a box, drop in a lure, and you never have to deal with installing bees. They will install themselves eventually.

25

10.

Times for Swarms

*W*e know swarms occur when bees are either crowded or triggered to reproduce. Both tend to occur at the same time: during honeyflows.

Here in the southern United States, we have our main swarming season from March through June.

In East Africa, the heavy swarming season takes place during the early and late monsoon season and again, six months later, during a strong honeyflow.

Take advantage of the winter season to get your traps ready.

Your location and micro-climate dictate when your local swarming season occurs. It should correspond with good honeyflows. To find out when your local swarm season is just ask the older beekeepers in your local club.

When to Build Your Traps

The absolute best time of year to build swarm traps is in winter. Winter is by far the most underused season for beekeepers. Use this time to get your equipment repaired for the next year, get old gear painted and then build out your swarm traps.

So many beekeepers get swamped in early spring doing things they should have done in winter. Don't be one of them.

When to Deploy Your Swarm Traps

Swarm traps should be set out the week before the earliest swarm has been seen in your area. Where I am, the earliest swarm I've seen is March 1, so I should have my traps deployed by the last week of February.

One month after the last known swarm for my area, I'll go pick up my traps. One year, however, I decided to leave them out all year. This turned out to be a bad idea. Bugs, ants, birds, and wasps can set up shop in the swarm trap, and you have not saved yourself any work. Plus, the biggest danger to your trap is a two-legged creature known as a teenager.

Bring the traps down after swarm season ends, and store them under a tarp until next year.

Ask a local expert when the swarming season is for your area.

How to Hang Swarm Traps

First, you really need to have a hefty nail to hang the trap on a tree. The trap masses maybe 3 kilos. But once a swarm sets up in it, you will have 1 to 2 kilos of bees who will quickly add a kilo of brood and a few kilos of honey. So guess what? The original nail you hung the trap on that worked at first is now over stressed, and the trap will collapse killing the hive. You don't want this to happen, so overcompensate and use a heavy nail. (Yes, this happened to me. *Please* learn my lesson without paying for it yourself.)

Use a big nail — new hives gain weight fast.

When you drive in the nail, remember that the trap will, being a dynamic structure, sway from time to time.

Drive it in at a 45-degree angle.

The design of a swarm trap has some true engineering beauty to it. Having the hang board on the back 12 inches above the center of gravity means it will act like a heavily dampened pendulum. In other words, if the trap gets hit, blown, or shaken it will self-correct back to proper orientation— Like a self righting boat.

This strategy requires that you hang the trap so it stays on the nail. Be sure to drive the nail into the tree at about a 45-degree angle and not to drive it in too far. You want a good length of nail sticking out to hang the box with some extra length of nail. The rocking motion tends to creep the trap up the nail a bit.

When you hang the trap, I'd recommend putting it 1 foot out of reach of a teenager. For example, if the box is two feet high, you should drive the nail in about 9 feet off the ground (or out of reach). In other words, the 6-foot height of a teenager + 2-foot height of the trap + 1 foot buffer space = 9 feet. If you plan to hang your boxes out of reach, you will need to bring a short step stool or a flipped 5 galleon bucket. (But remember to avoid ladders).

If you are on private land and teenagers are not a problem, I would hang the traps 6 feet off the ground. You will have simple, easy access, and it will be a breeze to check the traps.

Make sure that the boxes hang properly and are not lopsided, as the bees will always build the comb strait down in the direction of gravity. Here is a poorly hung box that will have wonky comb if it catches a swarm.

Don't hang traps wonky.

11.

Safety

Africanized Bees

In Texas, there is the chance of catching Africanized swarms. Also, there is always the chance to capture European honey bees that are vicious.

Either way, the swarms should be nice and well behaved; they don't get nasty till they grow in size. I wouldn't fret about what kind of bees you have until they build 6 to 8 frames of brood comb.

Once you know you have a hot hive, requeen it. Don't put up with aggressive hives unless you really enjoy getting the snot stung out of you.

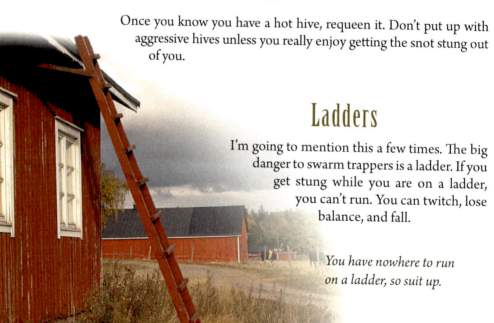

Ladders

I'm going to mention this a few times. The big danger to swarm trappers is a ladder. If you get stung while you are on a ladder, you can't run. You can twitch, lose balance, and fall.

You have nowhere to run on a ladder, so suit up.

Checking Swarm Traps

Don't have time to check your swarm traps daily? Well, daily checking is for the paper flowerpot style traps. You now have the best kind of swarm trap, and these traps start to pay off. First, you don't have to check them daily, or even weekly. Twice a month is okay. But you might catch a swarm and miss the next cast. Ideally, you want to check them every week

Personally, I check my traps every 10 days. It works for me, as I'm a busy businessman. I even have a spreadsheet of my traps. Below is my record system. I can record the trap number, if it is a veteran and my observations for the week.

Trap #	Veteran?	Description	05/06/11 (handwritten)	05/21/11 (handwritten)
1	no	Front yard	*Ss*	*Swarm*
2	yes	Logan Park Rd	*Na*	*Na*
3	yes	W of big oak	*Swarm*	*outyard*

The code I use is:

1. Na – no activity from scouts
2. Ss – some scouts
3. Swarm – a swarm is in trap
4. Outyard – trap is out at the outyard with a swarm, so don't check location

Pests

If you notice wasps coming out of the swarm trap, you need to take the trap down and remove the wasp nest. Don't let them stay or build up numbers. Don't use chemical spray, either as it will kill off any chance of catching a swarm. Leaving them also kills any chance of catching a swarm, as bees don't like wasps either.

Removal Methods

Have you caught a swarm? Congratulations! If you have made it this far, then you will likely be catching several swarms. That first swarm is something wonderful, like being kissed for the first time. That swarm is yours, through your work, planning, placement and perseverance.

So now what do you do? First, confirm you have a swarm. The only way I know I have a swarm and not a bunch of scouts is to watch the trap for a minute to see if any bees are bringing in pollen. I usually see foragers bringing in pollen in the first 30 seconds.

Scouts scout, foragers forage, but foragers never scout. Therefore, if a bee enters with a load of pollen on its legs, then there is a swarm in the trap.

I'll admit, I've taken down an empty box before thinking it had a swarm inside. If you do the same, you'll be in good company.

Second, after you confirm there is a swarm inside, then you play the game— the waiting game. Any early movement on that hive risks causing an exodus, and the swarm might take flight again. So you should wait a week or two for the swarm to build comb and the queen to lay eggs. Once there is brood in the comb, the swarm is anchored. It is not going anywhere and can be moved

You know you have caught a swarm when foragers bring in pollen (in the red circle above).

at that time. The two weeks is needed for a virgin queen as she has to mate before she starts laying eggs. I'll admit, you could take it down immediately, with minor risk.

Removing the trap is fairly easy if you built your swarm trap bee tight. Just wait until dusk when the last foragers are in, then plug the entrance with a paper towel. Take the trap down, and move it. You'll catch the majority or all the foragers, and you will still have light to see by. You could do this in the morning, but you have to show up in the dark.

If you didn't build the trap exactly bee tight, like mine, you need to prepare a bit more. You plug the hole, wait a couple of minutes to see if and where the bees are getting out and plug those holes with other bits of paper towel using your hive tool.

As a general rule, the swarm will still be docile. I move my swarm traps without a suit, but I'd recommend you suit up. If you still are nervous about bees buzzing you, you might drop the trap which will crash the fragile comb and possibly kill the queen, not to mention getting a few hundred bees angry.

Once the trap is at the apiary in its proper place, I pull the paper towel out of the hole and run. By the end of their journey, they are likely to be a tad angry and aggressive. Don't wait around. Come back the next day and check for any fallen combs, just as a best practice. If a comb has fallen, you can squish it back onto a frame because the combs are still so soft. You can transfer the hive from the trap to a hive body while you are there. This is also best practice as it allows you to put this trap back in service while it still smells of bees, raising the odds of catching another swarm considerably.

Do not leave the swarm trap sealed up with a hive inside out in the sun. The trap is not properly ventilated and the hive will likely die. I learned that the hard way.

12.

Using GPS

*I*f you are going to deploy quite a few traps, like more than five, I'd recommend using a GPS to mark the location of your traps.

More important, it marks the location **of the nails** you hung the traps on for next season. Trust me on this. You'll be sending me emails to thank me on that bit of wisdom. If you don't have a GPS, draw a quick map of all locations to remind you next year.

It is a good idea to label your GPS waypoints the same as your trap numbers.

GPS your swarm traps so you can find the nails next year.

13.
Legality

Trespassing

*T*respassing is going onto other people's land without permission.

Some countries or states have other laws that permit you to trespass legally. You should know the law on trespassing whether you swarm trap or not. Consult legal council if you don't.

Ownership

Generally, ask the landowner or tenant for permission to place swarm traps on their land. Be sure to ask permission to trespass to check the hives every few weeks. It helps to inform the landowner that these will decoy bees from establishing nests in their house walls.

The legal term for this is the right of ingress and egress from their land. This is if you wish to get a written statement.

Ask landowners for permission before hanging traps.

14.

Disease

*F*eral swarms, especially early ones, are usually cast off from the strongest, healthiest hives.

After all, sick hives don't build up strong to where they are the first to send off a swarm. Sick hives linger and often die.

Early swarms are usually sourced from the best survival genetics. Statistically, the donor hive resists disease, handles small hive beetle, and keeps the varroa mite in check. Now before you go tell me that might not be the case, I want to reiterate that I'm referring to a statistical mean. Out of 100 early swarms, the average genetics will be superior for disease resistance. This concept is yet another dissertation awaiting a grad student.

Obviously, an early swarm could be from a donor hive in a small cavity, or a hive that is being fed from an overzealous new beekeeper (Beekeeperous Noobeous, of which I was guilty myself). In general, you'll find that early swarms are healthiest. Covet them. Set them aside in the outyard to see if they are all that I say they are.

Later swarms can be anything, but in order for the swarm to have built numbers to swarm, you'd expect strong hives. Strong hives are not sick, statistically speaking.

15.

Location, Location, Location!

*L*ike fishing, swarm trapping requires care in picking where you prospect. This is where we leave science and move into true craft.

The craft of swarm trapping has some useful lore. As a general rule lore tends to show the patterns of experience outlined in rules of thumb.

Here is what we think we know from the lore of bait hives and swarm traps.

Sun Is Bad

This is a "Where not to put it" answer. Direct sun is bad. For some innate reason bees will reject traps that sit in full sun or part sun. Often, I've seen a trap rejected in part sun. I now put them all in the shade.

As an experiment, I've put identical traps 20 feet apart and the one in part sun was rejected every time. That taught me something. Learn from it.

Fence Lines

Bees navigate in many ways, using sunlight, using smell, and by landmarks. Fence lines are often used as a landmark and guide. So there is unusual bee traffic above them. Sort of like a river, and if you have swarm trap down below, they will smell it more easily as if it were a burger shack off a highway. The trick is

Bees use fence lines to navigate.

to hang traps along a fence line and keep them out of direct sun. That means you will want to find a tree along the fence line.

Use the shade.

Lone Trees

Do you have a nice big field with an old oak in the middle of it? That oak tree is where you want your trap, Somewhere where the trap is in shade.

Building Roofs

Lore has it that buildings are good landmarks for bees. This is likely referring to buildings in the country, where they act as prominent landmarks. So if you can hang a trap in the shade somewhere on the roof, you've got a good spot.

Look for a shady rooftop spot.

38

Tree Lines

Just like fence lines, tree lines act as highways for navigation. Place the trap a few feet into the tree line to keep it out of sun.

Tree lines are good places to find bees.

If bees swarm in a forest and no one is there, do they make a sound?

Deep in Forests

Rumor has it that forests are good spots. I've not had the chance to do this myself, as I've been hanging most of my traps in urban areas. It is quite reasonable that forests are bee country. As where there are trees, there are bees.

Urban Terrain

My most productive traps are in an urban area near a park. I hang a trap in my front yard, my backyard, and my side yard. Each season I catch a swarm in each at least once. Last year I got two in the front and two more in the back. Scientific studies have confirmed that the hive density in cities is higher

You should have your best swarm trapping success in urban terrain.

than the countryside. This is mainly because the cavity density is much higher, with all of those cracks and crevices in buildings. Add to that the better food sources with all the watered plants.

39

Abandoned Houses

If you are out in a farming area that has old abandoned houses, you will find they make good spots to hang a trap. They provide shade, a landmark, and a nice, easy-to-reach location. But beware the dangers of old houses: rotten floors, collapsing walls and the dangerously evil unmarked well outside. While, you could get hurt by falling through a floor or a wall falling on you, falling down a well will likely kill you. Take care.

This may be a good spot to hang a trap.

Adapt

If some locations aren't catching swarms, move your traps elsewhere. Some places just aren't as good as others. If there is an area that is good for you, double the number of swarm traps there. This test will determine if you are missing swarms because you don't have enough traps out.

Be aware of how weather affects the swarm rate. If your traps in the countryside don't catch many swarms during a bad drought year, don't assume the location is bad. If it is the conditions, you can't tell about that location until there is a normal year.

16.

Expectations

If you are like me, the question on your mind is, "If I do this, will it work?"

The answer is Yes!

If you hang swarm traps you will catch swarms!

As long as bees are known to visit plants in your area, you know there is a hive within three miles. That hive is likely to cast off a swarm every year, and more than likely there are other hives like it in the area.

If you hang your trap out of the sun, bait it with lemon grass oil and have patience, then your odds of catching a swarm are around 20% right off the bat. If the trap has old comb in it, or is a veteran trap, I'd put your odds for success at 40%.

Regional rainfall heavily influences swarming.

There is another factor: rain. Rain creates soil moisture, and soil moisture provides water to plants and puts more nectar in their blooms. More nectar means more food, hence more bees. That means swarms happen more often in higher rainfall areas.

Take for example the study called "Presence of Russian Honey Bee Genotypes in Swarms in Louisiana" that was put out in the March 2011 issue *Bee Culture*. The study was on bee genetics and conducted in Louisiana. The important tidbit in the study is that they used 200 swarm traps with lures to catch 147 swarms. That puts them at a 73% success rate. But that part of Louisiana is high rain, so I'll put 73% as an upper limit.

In Austin, Texas, in 2010, I had a 60% success rate, but some of my traps were veterans, plus we had an anomalous amount of spring rain. I would catch two swarms weeks apart in one box. Then I'd catch none in others—sort of like fishing.

You will catch swarms, no question about it. Be prepared to have a 30% success rate. Buy the extra hive bodies or find a friend who has empties that will loan you some. Don't be caught by surprise when you start catching swarms.

Transferring hives from a hanging box style swarm trap is made easy by frames.

17.

All Good Things Must End

The main swarm season is over by July here. That is the time to go collect all your swarm traps and put them in the shed, or store them as you wish.

Sometimes a rare fall swarm will find the traps and set up shop to your delight. Better feed that swarm, quick!

If you don't take the traps down, teenagers, birds, acrobat ants, wasps, etc., will find them. So take the traps down and store them. When winter comes around, get them ready again for spring deployment.

I once tried keeping them up year round to minimize my travel hassle. The next spring I found many smashed by kids or weather. It wasn't worth it to me, but you could try. I don't recommend it in North America. In equatorial regions, that might be a great idea.

The author next to a trap that caught a swarm a few weeks earlier. If you hang swarm traps, you will catch bees!

18.

GUTS!

This last bit is the most important. I've seen the statistics of educational e-books and at-home courses. The vast majority gather digital dust, being unused and unread. So this is your call to action.

GUTS stands for Go Use This Stuff! Deploying one swarm trap is better than thinking about deploying 10. Just go do it!

If you've made it this far, I'll let you know one last secret. You will catch swarms! Lots.

After you catch your first swarm, tell your friends, lecture to your club and spread the word! I appoint you a swarm trap ambassador.

Good luck, and good beekeeping.

19.

Advanced Topic: A Business Model

You can make a hobby business out of trapping.

I've seen offers of pest control renting swarm traps to apartment complexes to keep the bees from establishing in buildings. Some people charge around $50 per trap per month, which includes checking the trap weekly. This is something fun you can do and get paid for.

If you've ever done a cutout, you can get the landowner to help you. All you have to do is offer to hang a trap in a tree for free, or for rent, and tell the owner it will protect the house and nearby houses from bees establishing in them. This is true; it does act as a shield. Then give the homeowner a business card, preferably a magnetic one that sticks on their fridge, and have them call you if they see a lot of bees coming and going in the trap. This will work, as the owner feels he is empowered to protect his neighborhood with this little swarm trap. He'll tell his neighbors about it as well.

20.

Swarm Trapping
for Public Service

*S*warm traps will often keep bees from establishing hives in buildings.

Telling people this is the best way I know to get permission to hang hives. Also, it is a beekeeper's traditional duty to serve his community to minimize bad bee-human interactions. This is a great way to serve your community.

21.

Questions and Answers

Q: Will swarm traps make my hives swarm more often?

A: No. Hives will swarm because of in-hive reasons, not out-hive reasons. Your trap will be a good looking home to one of your swarms that comes out of a hive, however.

Q: Can I use a 5 gallon plastic bucket?

A: Sure. It just won't work worth a darn, however. It lets too much light through, it smells wrong, and you can't put frames in it without serious modifications. Combined, the odds of success are low, and the hassle factor high.

Q: What about using a 5 frame nuc?

A: The size is too small. You might catch a swarm, but you are below optimum size, and Africanized bees are known for going for smaller cavities than European bees. So you are increasing the odds of catching an undesirable swarm, and lowering your chances of catching any swarm.

Q: Bears are tearing into my successful swarm traps, what should I do?

A1: Check more often and take down successful swarm traps. I'd also hang them higher.

A2: Have you tried land mines and anti-tank guns? (Just kidding.)

Q: I've heard of making swarm traps out of cardboard forms for concrete, will this work?

A: Badly. They don't take frames well, and they aren't made of wood so they don't last. The hanging box style trap is superior in many regards.

Q: I think teenagers are taking down and smashing my traps, what should I do?"

A: Do some or all of these:

1. Camouflage the box with paint.
2. Change the location by a few dozen feet, picking a more hidden tree.
3. Hang the trap higher.

A2: See the A2 above for bears. (Again, just kidding.)

Q: Will swarm trapping make me more popular at parties?

A: NO! I've tried repeatedly. Learn from my mistakes. For details, ask my wife.

22.

Feedback!

Want to Be in My Upcoming Swarm Trapping DVD?

Have you built a swarm trap and caught bees?

Send me a photo! Tell me your statistics, like I had 10 traps and caught 9 swarms. I'll be taking some of the photos to put in the upcoming DVD on swarm trapping.

In your email, please state that "I, (your name), am giving McCartney Taylor all rights to use this photo in perpetuity." This is so I have the right to use the photo in the next DVD and future revisions.

When will I release this DVD? I'm hoping by December 2011. Check www.learningbeekeeping.com to see if it is ready.

Note: If you are a grad student in entomology looking for a thesis dissertation topic, I apologize for singling you out.

23.

Disclaimer

Although the author and publisher have made every effort to ensure that the information in this book was correct at press time, the author and publisher do not assume and hereby disclaim any liability to any party for any loss, damage, or disruption caused by errors or omissions, whether such errors or omissions result from negligence, accident, or any other cause.

Some people are severely allergic to bee stings, and the author and publisher recommend you consult your physician before attempting beekeeping.

Don't do anything dumb, risky, or any action that might get you hurt. The person in your mirror is the one responsible for your safety. Be smart, be safe.

Made in the USA
Middletown, DE
30 March 2015